IMAGES
of America

CHESAPEAKE BAY BRIDGE-TUNNEL

To Nina and Miguel,

9. 4. 15

The Chesapeake Bay Bridge-Tunnel opened on April 15, 1964, linking two Virginia communities that—even today—stand in stark contrast to one another. The Eastern Shore is much like it was 50 years ago, to the liking of its residents, many of whom have generational links to the peninsula. Virginia Beach, the southern terminus, is the most populous city in Virginia. Like its sister city, Norfolk, it has a transient population, many of whom serve at Hampton Roads' numerous military installations. This view shows the span from its southern terminus looking north in the months before the 1964 opening. At right, the South Toll Plaza building has not yet been completed. (Courtesy of the Chesapeake Bay Bridge-Tunnel District.)

ON THE COVER: The eventual namesake of the Chesapeake Bay Bridge-Tunnel, Lucius J. Kellam Jr., was a native of the Eastern Shore of Virginia and a 40-year member of the Chesapeake Bay Bridge-Tunnel Commission and its predecessors. Here, seated in the center of the convertible's rear seat, Kellam enjoys a breezy ride on the span during construction in the early 1960s. (Courtesy of the Chesapeake Bay Bridge-Tunnel District.)

IMAGES
of America

CHESAPEAKE BAY
BRIDGE-TUNNEL

John Warren
Foreword by Jeff Holland

ARCADIA
PUBLISHING

Published by Arcadia Publishing
Charleston, South Carolina

Printed in the United States of America

Library of Congress Control Number: 2015946010

For all general information, please contact Arcadia Publishing:
Telephone 843-853-2070
Fax 843-853-0044
E-mail sales@arcadiapublishing.com
For customer service and orders:
Toll-Free 1-888-313-2665

Visit us on the Internet at www.arcadiapublishing.com

Dedicated to the memory of Lucius J. Kellam Jr. and all others who believe they can change the world and then do something about it

CONTENTS

Foreword   6

Acknowledgments   7

Introduction   8

1.   The Ocean Highway   11

2.   Over and Under the Sea   27

3.   From Pines to Palms   81

4.   The Parallel Crossing   113

FOREWORD

The year 1964 was a remarkable one. The Beatles ruled the radio airwaves with nine solid hits, the Ford Motor Company unveiled the Mustang, and Martin Luther King Jr. received the Nobel Peace Prize.

In the midst of this historic year, the Chesapeake Bay Bridge-Tunnel, the largest engineering project in the world and one of mankind's greatest engineering achievements, opened to traffic on April 15. The opening of this water crossing completed the East Coast scenic shortcut to coastal destinations and capped an intensive eight-year project development, innovative design, and awe-inspiring construction period.

It was the end of the ferry era for the fleet of seven ferries that plied the waters of the Chesapeake Bay. As well, it was the beginning of a new era in transportation, with this engineering marvel providing a much safer and efficient crossing for the traveling public.

Lucius J. Kellam Jr.'s foresight, leadership, and dedication moved this project from a vision to the reality it is today. The facility was named and dedicated as the Lucius J. Kellam Jr. Bridge-Tunnel in 1987 in honor of his tireless efforts. It is only appropriate that when present and future generations think of the Chesapeake Bay Bridge-Tunnel, they remember Lucius J. Kellam Jr.

This graceful crossing continues to provide millions of travelers with breathtaking views of the mighty Atlantic Ocean and the Chesapeake Bay each year and includes a unique stopping point along the way. At the southernmost man-made island, customers can stretch their legs while enjoying fabulous views of the Chesapeake Bay as well as the unique local menu at the Chesapeake Grill and retail offerings at the Virginia Originals gift shop. Importantly, the Chesapeake Bay Bridge-Tunnel also boasts two visitor centers, one at North Plaza and the other at the southernmost man-made island.

Now the Chesapeake Bay Bridge-Tunnel is tunneling toward the future, as the project development of the Parallel Thimble Shoal Tunnel is underway, and the contract will be awarded in mid-2016 after procurement completion.

—Jeff Holland
Executive Director
Chesapeake Bay Bridge-Tunnel

ACKNOWLEDGMENTS

The story of the Chesapeake Bay Bridge-Tunnel rightly belongs to its keepers, the employees and commissioners of the Chesapeake Bay Bridge-Tunnel District. It is only with their enthusiastic support and their buy-in that this book came to be.

Paige Addison has worked for the bridge-tunnel throughout her adult life. She allowed me to pepper her with e-mails, phone calls (cellphone and desk phone), and visits. Her valued perspective and generosity are in evidence on every page of this project.

That is how the folks are at the bridge-tunnel. It is distinguished from other facilities in that it feels like it is owned by somebody and not something.

Throughout my work on the book, the bridge-tunnel's executive director, Jeff Holland, made it clear he felt the telling of this history was a top priority. He opened doors—and also drawers—and helped make connections whenever I asked.

At the *Eastern Shore News*, when I asked editor Ted Shockley if I could peek into his archives, he gave me everything. And the *News*'s crack photographer, Jay Diem, not only put dozens of photographs on disk for me, he even shot original images to boost the effort.

Bill Neville and Marion Naar with the Cape Charles Historical Society spent countless hours on several occasions combing through the museum's digital and print archives with me and helped me pull apart perfectly good displays to reach just the right photograph.

Finally, my thanks to Walter Grantz, Chesapeake Bay Bridge-Tunnel chief engineer during the Parallel Crossing. His expertise provided valuable insight into construction processes represented throughout the book.

INTRODUCTION

It must have been something like the notion of man walking on the moon.

One can imagine Lucius J. Kellam Jr. and his fellow true believers and the sideways glances they encountered when they first talked about crossing 18 miles of rough water with a fixed span where the Chesapeake Bay meets the Atlantic Ocean.

It was a fantastic idea at its core, that man could join two points of land so far apart.

The first English colonists made landfall at Cape Henry, only a few miles from the Chesapeake Bay Bridge-Tunnel's southern terminus. And as early as colonial days—when packet ships traveled between the lower tip of the Eastern Shore and Norfolk—men dreamed of a swift route linking the Eastern Shore to the Virginia mainland.

The first suggestion of the route was realized in 1884 by an engineer named A.J. Cassatt, who built the New York, Philadelphia & Norfolk Railroad, the NYP&N, which ran 95 miles from Delmar, Delaware, to the boomtown of Cape Charles, Virginia. There, at the Cape Charles Harbor, there were freight, passenger, and railcar ferry operations to the mainland of Virginia.

It was the NYP&N's successor, the Pennsylvania Railroad (PRR), that set the chain of events into play that would ultimately lead to the Virginia Ferry Corporation (VFC) and to the Chesapeake Bay Bridge-Tunnel.

The PRR was running three ferries—the *Pennsylvania*, *Maryland*, and *Virginia Lee*—to Old Point Comfort in Hampton and the Brooke Street Terminal in Norfolk. The three aged vessels were carrying passengers, freight, and some cars. The upstart Peninsula Ferry Company (PFC) saw an opening for automobile ferries.

From here, one needs a scorecard and perhaps an acronym guide.

The PFC started running car and passenger ferries between Norfolk and Cape Charles. The PRR, not to be outdone on its home turf, responded by running ferries to Little Creek in Princess Anne County (present-day Virginia Beach). PFC asserted that PRR could not legally run automobile ferries and filed a suit. In August 1930, PRR responded by forming the Virginia Ferry Corporation (VFC). Later that year, PFC was hard at work building a new terminal at Cape Charles. But, by 1933, PFC dropped its suit, accepting substantial stock in VFC. It was now, finally, a profitable venture, and on April 1, 1933, a 26-mile run between Cape Charles and Little Creek began. The numbers of ships and vehicles carried increased steadily.

Cape Charles, a boomtown 12 miles north of the southern tip of the Eastern Shore, had first seen its fortunes buoyed by the railroad and then by becoming a hub for vehicular ferry service.

It was so crowded in town that locals walking to the beach had to press against the buildings on Front Street, the main drag. Coming back from the beach, they had to walk in the streets. During the holidays, people would park their cars well outside of the town limits and hike into town.

It was a melting pot of Germans, Spaniards, Italians, Chinese, and Lithuanians. There were guys with nicknames like Bootie and Sly. When the ferries came in, the local hangouts were bursting at the seams. There were 20 to 30 arrests in a night sometimes.

In 1935, the Ocean Highway Association, a cooperative that included business owners, the Virginia Ferry Corporation, and the American Automobile Association, was formed to promote north-south travel using US 13 from New York to Norfolk, Virginia, then US 17 to Jacksonville, Florida. Its slogan was "from Pines to Palms." In the day before interstates, its main selling point over US 1 was "Avoid all the traffic in Philadelphia and Washington." Of course, the Ocean Highway was heavily dependent on the Cape Charles ferries.

During World War II, Cape Charles was busier than ever, with PRR steamers carrying railcars headed to Norfolk and VFC ferries occupying its harbor.

It is hard to imagine anyone was anxious to see an end come to the romance of the ferry days, of meandering yet restful 90-minute boat trips across the Chesapeake Bay. "The world's champion pinch-penny cruise," one writer called it.

But, by 1953, change was afoot. VFC shifted all passenger and vehicle traffic from Cape Charles to near the southern terminus of Virginia's Eastern Shore, in Kiptopeke. There were five vessels handling an average of 1,800 cars and 7,000 passengers on a daily basis, and the numbers were climbing.

In 1954, the Chesapeake Bay Ferry District was created by the state with the intent of restoring Cape Charles–Old Point Comfort ferry service. On May 17, 1956, the VFC ceased to exist; the Chesapeake Bay Ferry District bought the company for $13 million. With Hampton Roads growing and increased East Coast travel and interstate commerce, the Little Creek ferry system quickly became the busiest in the world. Its fleet of vessels was making 90 one-way crossings per day. But it could not keep up with the crossing demand.

The ferry district's true charge was soon enough obvious. It would not be restarting ferry service between Cape Charles and Old Point Comfort. The case had long been building for a different solution, one long thought impractical, even unimaginable.

In 1956, the Virginia General Assembly cracked the door open just wide enough for Lucius J. Kellam Jr. and his army of true believers. It passed legislation for an engineering study, effectively challenging Kellam and his ferry district to prove the fixed crossing was feasible.

It was the opportunity Kellam—the ferry commission chairman—had been biding his time for since at least since 1950, when the Bay Bridge opened, linking the eastern and western shores of Maryland.

When men dared to dream of a span crossing the Chesapeake Bay, if the sheer scope of the ambition did not stop such talk, if the cost of the project did not stop it, then the idea of blocking two of the world's busiest shipping channels—including one relied upon by the US Navy—surely could.

The study by Sverdrup & Parcel, Inc., of St. Louis, Missouri, suggested a solution: a series of bridges and tunnels, tied together by miles and miles of low-level trestle and four man-made islands. Wilbur Smith & Associates, traffic consultants, concurred. On October 24, 1957, following a public hearing, the Virginia State Highway Commission upheld the Chesapeake Bay Bridge and Tunnel Commission's proposed north-south route.

The propaganda campaign along the Eastern Seaboard led with a mesmerizing notion: "Over and Under the Sea."

It was a practical solution but not an easy one. In the coming years, the problems of financing and building the span often felt insurmountable.

The largest bond issue in state history—$200 million, involving no state money—was negotiated. On September 7, 1960, revenue bonds were turned over to investment bankers, and the orders were issued: build the bridge-tunnel. A joint venture of four firms was called on to design and build the span. They included Tidewater Construction Corporation, Merritt-Chapman & Scott Corporation, Raymond International, Inc., and Peter Kiewit Sons' Company.

The story of its construction—the obstacles faced in building in the midst of a raging Chesapeake Bay, many miles from land—would inspire media coverage around the world.

The project would rely on three custom-made machine monsters: the *Big D* pile driver, the *Two-Headed Monster*, and the *Slab Setter*. Four man-made islands would rise in the middle of the

Bay. Two mile-long tunnels would be lowered into trenches, and two high-rise bridges and 12 miles of low trestle would be built.

On April 15, 1964, the grand ferries were idled, dispatched to a new route that would connect Cape Lewes, Delaware, to Cape May, New Jersey. With the release of 50 seagulls, a procession of cars and buses was unleashed, and the Chesapeake Bay Bridge-Tunnel was open.

Along the Eastern Seaboard, Madison Avenue had made it easy to find the bridge-tunnel, with familiar signs featuring a white seagull against a blue background. "Follow the sign of the soaring gull," the advertisements read. And motorists did.

The American Society of Civil Engineers declared it one of the Seven Engineering Wonders of the Modern World. *Readers Digest* went a step further, calling it one of the Five Future Wonders of the World.

It is something much more than a convenience, a means of shaving 75 miles from automobile travel between Wilmington, Delaware, and Virginia Beach, Virginia. The bridge-tunnel is an anomaly in the world of highway infrastructure, a link between destinations that became a destination in its own right.

By nature of its political designation—as a district with its own taxing authority—the Chesapeake Bay Bridge-Tunnel is an independent, self-sufficient city in the middle of the bay (it even has its own police force). Its first northbound island is a must stopover for travelers, with a restaurant, gift shop, and fishing pier.

The Chesapeake Bay Bridge-Tunnel links two distant, curiously disparate expanses of tidal land that would never have been joined but for the determination of men who were enraptured by the notion of "What if?" Even those who worked tirelessly to make it happen marveled at the accomplishment.

"For all practical purposes, the bridge-tunnel is situated in the middle of the Atlantic Ocean," said Percy Z. Michener, the engineer who devised the idea of the "over and under" span.

In 1964, man conquered the Chesapeake Bay. The moon would have to wait a few more years.

One

THE OCEAN HIGHWAY

Capes Charles, Virginia, became a boomtown in 1884 when a rail line was completed that ran 95 miles, from Delmar, Delaware, through the Eastern Shore of Maryland and Virginia, to Cape Charles, 12 miles above the southernmost point of Virginia's Eastern Shore. The railroad was called the New York, Philadelphia & Norfolk Railroad, or the NYP&N. The freight, passenger, and railcar ferry operations to the mainland of Virginia were built at the Cape Charles Harbor, shown here in 1928 with the ferry *Maryland* and the tug *Wicomico*. (Courtesy of the Cape Charles Historical Society.)

In those days, Cape Charles's bustling harbor meant a melting pot of many nationalities and streets full of vendors, buses, and taxis. It was the NYP&N's successor, the Pennsylvania Railroad (PRR), that set the chain of events into play that would ultimately lead to the Virginia Ferry Corporation, which in turn set the stage for the Chesapeake Bay Bridge-Tunnel. As far back as 1885, car floats (below), or barges, transported railroad cars 36 miles across the Chesapeake Bay from Cape Charles to Port Norfolk, in Portsmouth. By 1929, the cars were shipped from Cape Charles to Little Creek Inlet, where the cars were shifted to the railroad. The shift cut 10 miles off the route. The float operation continues today, one of only two remaining on the East Coast. (Both, courtesy of the Cape Charles Historical Society.)

By 1930, the Pennsylvania Railroad was running three screw-propelled steam ferries, *Pennsylvania*, *Maryland*, and *Virginia Lee*, carrying passengers, freight, and some cars to Old Point Comfort in Hampton and to its Brooke Street terminal in Norfolk. All three vessels were old. *Maryland*, pictured below, was built in 1907. Pictured above are passengers unloading from *Maryland* in the 1940s. During World War II, people came to the Eastern Shore from the Bahamas to help harvest potatoes. The Bahamians went to Old Point Comfort in Hampton first to get physicals and paperwork taken care of. After the harvest, the Bahamians would travel to Pennsylvania to work there. (Both, courtesy of the Cape Charles Historical Society.)

One of the early steam-powered ferries on the Eastern Shore was the *Virginia Lee*, shown at dock in Norfolk. *Virginia Lee*, built in 1928, was named for the daughter of the PRR president. She had lush passenger cabins and elegantly furnished dining facilities. After many years of service on the Cape Charles–Norfolk route and a tour in Brazil on the Amazon River, *Virginia Lee* returned to service in the United States in 1948 as the ferry *Accomac*. She was damaged by fire in 1964 and scrapped and towed, becoming part of the Ghost Fleet of Mallows. Her remains are still in the shallow water of Mallows Bay, on the Maryland side of the Potomac River. (Courtesy of the Cape Charles Historical Society.)

In 1930, it was war. The Peninsula Ferry Company (PFC) was formed to operate car and passenger ferries between Norfolk and Cape Charles. The Pennsylvania Railroad (PRR) responded by running ferries to Little Creek in Princess Anne County (now Virginia Beach). PFC asserted that PRR could not run ferries for non-railcars. The PRR responded by forming the Virginia Ferry Corporation (VFC). (Courtesy of the Cape Charles Historical Society.)

In 1931, competition was fierce. The PRR started making openings on the freight decks larger to provide more space for cars. The PFC added the *Hercules*, which ran with the *Pioneer*. But the PFC vessels were not designed for the open bay and experienced mechanical problems and grounding in Cape Charles. The PRR vessels, for their part, could only handle cars, not trucks. In 1933, the PFC dropped its lawsuits again the VFC and received cash and part ownership in the VFC. Above, automobiles prepare to load onto a ferry at Old Point Comfort in Hampton. (Both, courtesy of the Cape Charles Historical Society.)

With the beginning of World War II, Cape Charles Harbor was overrun with arrivals and departures. They included the PRR steamers to Norfolk, PRR car floats, and VFC ferries. Things got more dire on July 9, 1942, when the *Virginia Lee* was requisitioned for war services. With the *Pennsylvania* having been scrapped in 1940, that left only the *Maryland* on the Norfolk route. In March 1944, the *Richard Peck*, a former Long Island Sound night boat originally built in 1892 by Harlan and Hollingsworth of Wilmington, Delaware, found new life when she was purchased by the PRR. (Courtesy of the Cape Charles Historical Society.)

The *Richard Peck* was brought south and refurbished by the VFC and renamed the *Elisha Lee*. The boat had a storied history, having been designed by the American marine artist Archibald Cary Smith, who also designed the 1881 America's Cup defender *Mischief*. She served as barracks for construction workers during construction of Naval Station Argentia, Newfoundland, from 1941 to 1943. Beginning in February 1943, she provided electrical power in Newfoundland. The *Richard Peck* was decommissioned by the Navy in December 1943 in Norfolk, Virginia, and transferred to the Virginia Ferry Corporation. (Both, courtesy of the Cape Charles Historical Society.)

Elisha Lee is pictured here at Old Point Comfort. In 1953, service was discontinued between the Eastern Shore and Old Point Comfort. The act that would set into motion the mechanism to create the bridge-tunnel was, in fact, a reaction to that discontinuation of service. When the ferry district was created, it was to restore service to Old Point. Not only did that not happen, but, in 1957, another automobile ferry service to Old Point Comfort—from Willoughby Spit—was replaced by the new Hampton Roads Bridge-Tunnel. The Norfolk-Portsmouth Bridge-Tunnel (now known as the Midtown) had been built in 1952. (Courtesy of the Cape Charles Historical Society.)

The *Elisha Lee* transported record numbers of passengers between Norfolk and Cape Charles. With two triple-expansion steam engines, she was one of the fastest ferries on the lower Chesapeake Bay. She was 303 feet long and could accommodate 1,200 passengers. When refurbished, above, she could carry 75 cars. Below, dining accommodations were standard on the modern ferries given the 90-minute passage. What would fare have been like? A typical dinner would be Yankee bean soup, a roll and butter, chocolate pudding, and coffee or tea. The price? 30¢. One could substitute Virginia beef stew for the soup for an additional 5¢. (Both, courtesy of the Cape Charles Historical Society.)

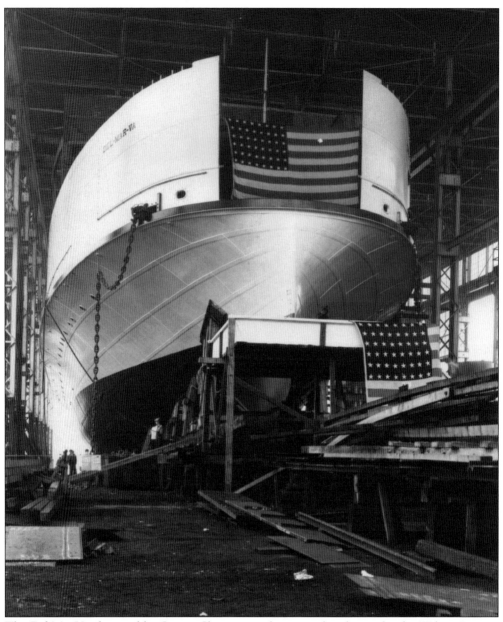

The *Del-Mar-Va*, designed by George Sharp, consulting naval architect for the Wilson Line, is pictured here on November 2, 1933, just prior to her launch at the Pusey and Jones Company in Wilmington, Delaware. Her keel was laid on June 22, 1933, and she was launched on November 22, 1933. The *Del-Mar-Va* was capable of carrying 80 automobiles and could also accommodate buses. Her bow and stern doors could be closed in bad weather. The passenger deck featured amenities, including a restaurant and snack bar as well as a ballroom and passenger lounges. Below deck, there were bunks and showers for truckers. The engines were reliable, easy to operate, and responded quickly to bridge commands. The ferry made her first crossing from Cape Charles to Little Creek on January 7, 1934. By 1955, when demand had reached fever pitch, the *Del-Mar-Va* was lengthened by 90 feet, at a cost of $750,000, so that she could accommodate 104 cars. (Courtesy of the Cape Charles Historical Society.)

When vehicle transportation doubled across the Bay in 1933 and 1934, a contract was awarded to build a second new ferry, the *Princess Anne*, named after Princess Anne County in Virginia, the present-day City of Virginia Beach. Renowned French American industrial designer Raymond Loewy designed the ship. Loewy, once featured on the cover of *Time* magazine, had been called "the man who shaped America." He designed everything from cigarette packs and refrigerators to cars and spacecraft—and ships. Loewy lived by his own famous MAYA principle: "Most Advanced Yet Acceptable." The *Princess Anne* launched to much fanfare on May 18, 1936. She had the same motor as *Del-Mar-Va* and the same carrying capacity for motor vehicles. But her center overhead was higher and rounded and gave the *Princess Anne* the appearance of a much larger vessel. On July 10, she made her maiden voyage from Cape Charles to Little Creek. (Both, courtesy of the Cape Charles Historical Society.)

In early 1946, the *Princess Anne* entered Newport News Shipbuilding for modifications. It removed prominent features of the Loewy design; the rounded sheet metal had been determined to be impractical. By the early 1950s, VFC vessels were carrying an average of 1,800 cars and 7,000 passengers daily. In 1954, the *Princess Anne* was lengthened to accommodate growing traffic. At Maryland Shipbuilding and Drydock Company in Baltimore, she was cut vertically, with 90 feet added. This increased the automobile capacity from 80 to 104. She returned to service in April 1954, after only 75 days. At 350 feet long, the *Princess Anne* was the longest in the ferry fleet. After her tenure with the VFC, the *Princess Anne* served a ferry operation between Cape May, New Jersey, and Lewes, Delaware. In 1993, she was sunk off Fishers Island in Palm Beach County, Florida, becoming a large artificial reef for fishermen and scuba-diving enthusiasts. (Both, courtesy of the Cape Charles Historical Society.)

The interior of the *Princess Anne* included amenities such as a lunch counter, a dining room, and a ballroom/lounge. Advertising literature of the period summoned passengers with the slogan, "Relax . . . Dine . . . Save Time." A Virginia Ferry Corporation brochure read, "Relax for an hour and a half on big bayliners away from the wheel. While you're steaming, enjoy a snack or delicious meal in the dining room. Stretch out in the comfortable picture window lounges—or stroll on the deck of one of the seven big 1,200-passenger ships. . . . This is the way to drive North-South and enjoy your trip." (Courtesy of the Cape Charles Historical Society.)

By 1950, Virginia Ferry Corporation's *Maryland* had been scrapped, and only the *Elisha Lee* (pictured) remained on the Cape Charles–to–Norfolk run. *Elisha Lee* made her last run on February 28, 1953, from Cape Charles to Norfolk—the day before she was expected to fail a rigorous Coast Guard inspection. This was the end of passenger/freight steamboat operations in Cape Charles and the end of the Norfolk–Old Point Comfort–Cape Charles route. Only railroad-car floats still operated out of Cape Charles. *Elisha Lee* was scrapped in Baltimore in 1953, ending 61 years in service. (Courtesy of the Cape Charles Historical Society.)

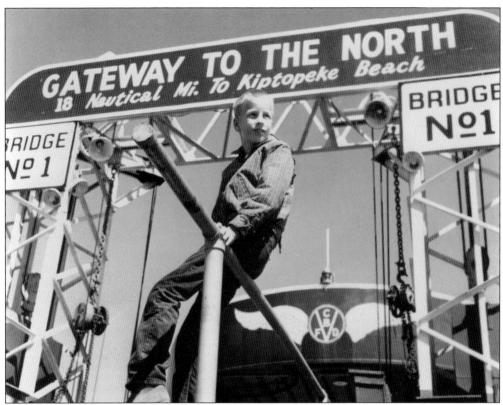

In 1950, The VFC opened its new terminal at Kiptopeke Beach, eight miles south of Cape Charles. By 1953, the VFC handled all passenger and vehicle traffic from Kiptopeke. There were five vessels handling an average of 1,800 cars and 7,000 passengers on a daily basis, and the numbers were climbing. On May 17, 1956, the VFC ceased to exist when the Chesapeake Bay Ferry District Commission, a state government agency, bought the company for $13 million. It continued to operate until the Chesapeake Bay Bridge-Tunnel opened in 1964. The demand had built over many years toward a solution for north-south passage long thought impossible—a fixed span linking the Eastern Shore with the Virginia mainland. With the closing of the Kiptopeke Ferry Terminal (below) in 1964, the ferries were all transferred to the Lewes, Delaware, to Cape May, New Jersey, run. (Both, courtesy of the Cape Charles Historical Society.)

More than 50 years after ferry service discontinued from Kiptopeke, a reminder of the terminal remains in the form of nine World War II–era concrete merchant ships. In December 1948, the ships were sunk to form a breakwater from the tidal action and severe weather of the Chesapeake Bay. Just like its wood and steel counterparts, a concrete ship will float as long as the weight of water it displaces is more than its own weight. In the case of a ship, much of its interior space is empty, and the density of the entire ship is less than the water, allowing it to float. In 1942, two dozen merchant ships were commissioned of concrete or ferro-cement. Nicknamed "the floating skyscrapers," the fleet also came to be known as McCloskey ships, after their contractor, Matthew H. McCloskey and Company of Philadelphia. (Photographs by Jay Diem.)

The disappearance of the ferries in the 1950s took a sudden and dramatic toll on the town of Cape Charles. Reminders of the bayside town's heyday remain in this photograph, taken in the 1970s on Cape Charles's main drag, Mason Avenue. The words "Wilson's Department Store" can be seen on a building, by then converted to apartments. Mason Avenue was once resplendent with retail stores, including the likes of department stores such as McCrory's and Brown's. There was also an appliance store, multiple drugstores, hotels, grocery stores, barbershops, a roller-skating rink, and a movie theater. There were six doctors in the town at one point. By the 1970s, little of that remained. The population was more retirees and widows. Businesses closed—even the school closed. Historic houses could be bought for little more than the cost of a new automobile. But the town has enjoyed a resurgence in the 2000s with the construction of a golf community abutting its borders and a healthy abundance of boutique stores—and shoppers—on Mason Avenue. (Courtesy of the Cape Charles Historical Society.)

Two

Over and Under the Sea

In 1956, the scales had tipped. The Virginia General Assembly passed legislation for an engineering study. By 1960, the Chesapeake Bay Bridge and Tunnel Commission signed a purchase contract. This 1960 photograph shows commission chairman Lucius J. Kellam, second from left, with Gen. Leif Sverdrup (left), Sherwood E. Liles (second from right) of Tidewater Construction, and Edward R. MacKethan (right), a trust officer for Virginian National Bank, who helped arrange the bond issues for the bridge-tunnel. (Courtesy of the Cape Charles Historical Society.)

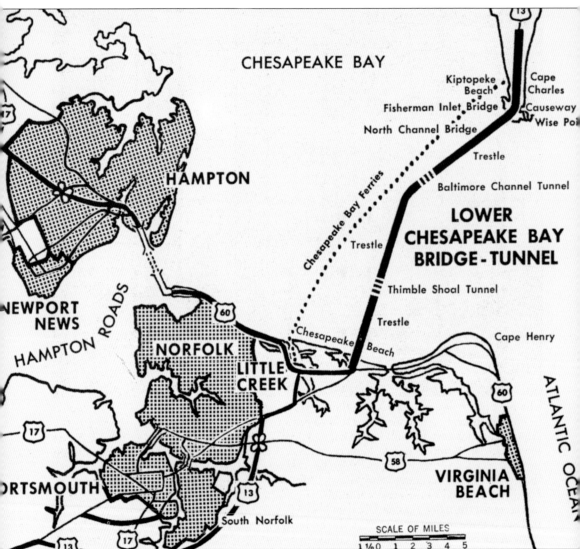

The engineering study commissioned by the Virginia General Assembly was performed by Sverdrup & Parcel, Inc., of St. Louis, Missouri. The study found the long-discussed crossing was feasible and recommended a series of bridges and tunnels following the path of the invisible line that separates the Chesapeake Bay from the Atlantic Ocean. Wilbur Smith & Associates, traffic consultants, concurred following hundreds of interviews, origin-destination studies, and field studies. Another route was considered, an 18.9-mile east-west route that would have linked Cape Charles with a spot just north of Buckroe Beach in Hampton. But studies showed the Fisherman Island–Virginia Beach link would draw up to 50 percent more interest from travelers. On October 24, 1957, the Virginia State Highway Commission concurred with the bridge-tunnel commission's recommendation in favor of the north-south route. (Courtesy of the Chesapeake Bay Bridge-Tunnel District.)

The next phase was raising $200 million toward construction of the bridge-tunnel. It would be entirely funded by revenue bonds sold to private investors, with toll money used to pay the principal and interest. It would be the largest bond issue in Virginia's history. The other business that needed a resolution was deciding on a name. It took nearly all of 1961 for the commission to decide between Virginia Capes Bridge-Tunnel, Virginia Capes Crossover, and various names that incorporated "oceanway" and "seaway." Ultimately, the group decided to go with the name that been attached to the project from the start, Chesapeake Bay Bridge-Tunnel. "The public had accepted the name," Lucius J. Kellam Jr. said in October 1961. Among its promotional efforts (below), a three-dimensional scale of the bridge-tunnel was produced (above) that travelled to many major cities. The model is today on display at the Cape Charles Historical Society Museum. (Both, courtesy of the Chesapeake Bay Bridge-Tunnel District.)

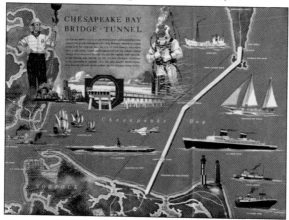

BAYSHORE CONCRETE PRODUCTS

CORPORATION

CAPE CHARLES · VIRGINIA

BCP

PRECAST AND PRESTRESSED
STRUCTURAL CONCRETE COMPONENTS
FOR CONSTRUCTION

CATALOG BC-1

Consulting engineers Sverdrup & Parcel were retained to design and supervise construction, and Tidewater Construction Corporation was the general contractor in a joint venture of four firms that also included Merritt-Chapman & Scott, Raymond International, and Peter Kiewit Sons' Company. The vast amount of concrete that would be required for the project necessitated the creation of a business that persists even to this day—Bayshore Concrete. Bayshore would be responsible for precast and prestressed piles, piling caps, decks, and other prefabricated concrete components. The company was built in Cape Charles for maximum convenience for bridge-tunnel construction. This situated it well not only for marine access but also for rail deliveries. (Courtesy of the Chesapeake Bay Bridge-Tunnel District.)

At Bayshore Concrete, steel cables joined pile sections together to create long piles for driving into the Chesapeake Bay's floor. The cylindrical piles that formed the trestle sections were made by a unique process called "Cen-Vi-Ro," and were cast in 4-, 8-, 12- and 16-foot lengths. The patented method involved centrifugal force, vibration, and rolling. Workers inserted welded cages into a cylindrical steel form, after which concrete was introduced as it spun at more than 60 revolutions per minute. The force from the spinning compressed the concrete. Among other things, the patented process meant a dryer concrete mix could be used, which meant a more durable end product. Each section—16 feet long and 54 inches in diameter—then stood vertically to cure for 28 days. The cured sections were aligned on a horizontal bed to produce the desired pile length. The cables were then fed through the piles and tensioned using hydraulic jacks to withstand high stress. The end result was "prestressed" piles at the desired lengths. (Above, courtesy of the *Eastern Shore News*; right, courtesy of the Chesapeake Bay Bridge-Tunnel District.)

The $3.5-million casting yard, although set up to speed bridge-tunnel construction, was designed to become a permanent facility. The company possessed a high degree of mechanization even at its outset. Piles varied in length from 64 to 172 feet. The completed pile was picked up by a crane and loaded onto a barge to move to the job site. (Both, courtesy of the Chesapeake Bay Bridge-Tunnel District.)

The plant built more than 2,500 pilings for the project and emerged from the 1961–1964 projects as one of the top sources of marine-grade precast/prestressed concrete structures. Prestressing is a method of applying tension to concrete to increase its structural strength. Though Bayshore Concrete was not included in the team for the Parallel Crossing in the 1990s, it continues to work on high-profile projects today, including the new Tappan Zee Bridge project over the Hudson River in New York and the Great Egg Harbor Bridge in New Jersey. These projects mean hundreds of new jobs for the Cape Charles company. (Both, courtesy of the *Eastern Shore News*.)

A trio of "mechanized monsters" was responsible for the construction of the bridge-tunnel's trestles. The first of these was the $1.5-million pile-driving barge *Big D*. It was specially designed to cope with the heavy waves and winds at the entrance to the bay. The pile driver—70 feet wide by 150 feet long, then the largest in the world—appears in this photograph to be floating in the Bay. In fact, *Big D* was supported on four 100-foot-long steel-pipe legs (or "spuds"). Each leg, six feet in diameter, was equipped with a "shoe," a pontoon 28 feet in diameter, to reduce penetration into the seabed. The legs could be raised or lowered into the seabed by 500-ton-capacity air jacks. This provided a platform that was both fixed and also mobile. (Courtesy of the Chesapeake Bay Bridge-Tunnel.)

One set of three piles is called a bent. The two outer piles are angled for added stability, and every 75 feet, a row of piles supports the trestle. To drive a pile, *Big D*'s crane, with a 185-foot boom, would lift piles from the barge then swing the upended pile into a guiding cradle for driving. Water jets would then give it a start by sinking it to the soft bottom, after which a steam hammer would drive the pile home. The crane would pick up piles weighing more than 150,000 pounds. This pile was driven off Wise Point in May 1962. (Courtesy of the Chesapeake Bay Bridge-Tunnel District.)

The prestressed concrete cylinder piles used for the bridge-tunnel were developed by Raymond International, Inc., one of the project's contractors. The piles could not vary more than three inches in any direction from the prescribed location. The photograph above shows the first of more than 2,500 piles being driven. Below, the final pile is being driven. The pile driver would either deliver a set number of blows or drive the pile to a predetermined depth. (Both, courtesy of the Chesapeake Bay Bridge-Tunnel District.)

Children play in the sand near the southern, Chesapeake Beach, end of the bridge-tunnel. In the distance is the second of the three mechanical titans that built the Chesapeake Bay Bridge-Tunnel: the *Two-Headed Monster*. The *Two-Headed Monster*'s intent was to trim the pilings on one side, and, on its other end, to bridge the set of pilings with a heavy beam called a "pile cap." This completed combination is called a "bent" and serves to support the precast deck sections of the trestle at 75-foot intervals. Visible on the deck of the *Two-Headed Monster* is a derrick—a style of crane with a pivoting arm. (Courtesy of the Chesapeake Bay Bridge-Tunnel District.)

The *Two-Headed Monster* was a 255-foot-long bridge that moved from pile cap to pile cap on its own rail tracks. It had a derrick on each end. One derrick carried the mechanism for cutting off the piles, while the other derrick capped the piles that had just been cut. Each pile cap weighs 30 to 35 tons. (Above, courtesy of the *Eastern Shore News*; below, courtesy of the Chesapeake Bay Bridge-Tunnel District.)

Prior to placing pile caps on the sets of three pilings, workers had to cut off the piles. They accomplished this with jackhammers mounted on a track, above. Below, concrete is poured through holes in the pile caps into the hollow piles. After the piles were driven, they were filled with sand, then concrete for the top four feet. The top four feet also included precast steel. The concrete-and-steel-form combination provided a strong structural connection between the trio of piles and the pile caps and also helped buttress the five-inch-thick wall against the potential impact of a boat or ice. (Above, courtesy of the *Eastern Shore News*; below, courtesy of the Chesapeake Bay Bridge-Tunnel District.)

Following the *Big D*, above, the *Two-Headed Monster*, left, travelled across cylinder piles, cutting the piles to appropriate elevations and then placing pile caps in preparation for the roadway slabs. Piles varied from 64 to 172 feet long, with an average length of 100 feet. The depth was determined from soil borings taken along the bridge-tunnel's planned route. Each pile was designed to carry 160-ton loads. The trestles account for nearly two-thirds of the Chesapeake Bay Bridge-Tunnel's span and include piles, pile caps, and 75-foot-long prestressed deck units. The trestle sections required more than 4.7 million cubic feet of concrete and 29 million pounds of steel. (Both, courtesy of the *Eastern Shore News*.)

Nine survey towers were built, about two miles apart, from which the bridge-tunnel's precise route was determined. These towers were needed to control all construction activities across the wide crossing of the Bay. They were vital for pile driving, trestle alignment, and tunnel placement, among other things. Total deviation on the bridge-tunnel is less than half an inch, and that accuracy was assured not by GPS—a technology that did not exist in the 1960s—but by telescope, compass, and triangulation from the nine survey towers. (Both, courtesy of the Chesapeake Bay Bridge-Tunnel District.)

Above, workmen stand on a section of highway in the middle of the Chesapeake Bay, dwarfed by the *Slab Setter*. The *Slab Setter* was the last of the three massive machines in the construction procession, specially designed to assemble the concrete trestle portions of the Chesapeake Bay Bridge-Tunnel. The *Slab Setter* moved in 75-foot strides, placing the four sections of roadway on top of the pile caps. Then it would pick itself up and wheel itself forward, sideways, putting it in place to lay the next four prestessed sections for the next span. Each section is 75 feet long and weighs 75 tons. (Both, courtesy of the Chesapeake Bay Bridge-Tunnel District.)

The low-level trestle is formed of prestressed concrete deck-girder slabs, laid four abreast. There are 825 precast prestressed concrete spans. Each slab is 75 feet long and weighs about 75 tons. The four girders are bound together to create a single deck span. Below, workers guide a slab into place as it is lowered by a crane operator. (Both, courtesy of the Chesapeake Bay Bridge-Tunnel District.)

Here, the approach spans for the high-clearance North Channel Bridge are constructed with cast-in-place concrete. The rails are used to guide a device called a "screed," which planes the wet concrete into a smooth surface. Below, the *Big D* pile driver can be seen in the background. The roads on the original span are a full 10 feet more narrow than the Parallel Crossing's roads would be. The original span is also minus the shoulders on each side that were installed with the Parallel Crossing project in the 1990s. (Both, courtesy of the Chesapeake Bay Bridge-Tunnel District.)

Three ready-mix trucks—the tanker of the third is visible in the background—are discharging at the same time while pouring a base slab. The trucks were working on a cofferdam on one of the islands. A cofferdam is a temporary retaining structure whose function is to restrain earth and water from an excavation. The trucks belong to Southern Materials Company of Norfolk, which was awarded the contract for supplying rock, sand, and gravel for the bridge-tunnel project. The company in turn negotiated contracts with three other companies in Virginia, Vulcan Materials, Trego Stone Corporation, and Tidewater Crushed Stone Company, and one in North Carolina, Superior Stone Company. (Courtesy of the Chesapeake Bay Bridge-Tunnel District.)

Finishing steps for the roadway included the mounting of aluminum cable trays to the outside of the trestles and bridges. The trays hold utility lines and cables, providing easy installation and ready maintenance access. Handrails, lighting, telephones, and asphalt completed the trestle. The completed roadway had 18-inch safety walks on each side and a curb-to-curb width of 28 feet. (Both, courtesy of the Chesapeake Bay Bridge-Tunnel District.)

E.C. Ernst, Inc., below, was among the electrical contractors for the bridge-tunnel project. The company, founded in 1915, has counted among its clients the White House, the J. Edgar Hoover Building, the Kennedy Center, Epcot Center at Disney World, and Washington National Cathedral. While there were four firms in what was called a joint venture, all told, more than 800 subcontractors and suppliers had a part in the construction and completion of the Chesapeake Bay Bridge-Tunnel. (Both, courtesy of the Chesapeake Bay Bridge-Tunnel District.)

More than 2,000 people worked on the bridge-tunnel construction, employed by 800 subcontractors. The bridge-tunnel's workers came from a myriad of backgrounds and locations. Some were experienced laborers, tradesmen, and engineers who chased jobs from state to state. Others were locals, still in their teens and living at home. Above, Bill Sanders, who worked on the *Big D* pile driver, stands in the middle of a group of coworkers, a pipe wrench slung over his shoulder. For Sanders, as for many who worked on the bridge-tunnel, the project was a defining experience—one he would share with people the rest of his life. (Both, courtesy of Donna Sanders Corbus.)

Above, brothers S.E. "Sam" Liles (left) and J.R. "Dick" Liles (right) of Tidewater Construction inspect bridge progress. Sam founded Tidewater and was president of the company during the bridge-tunnel project. Dick Liles was a project manager and vice president of Tidewater Construction and went on to succeed Sam as president, retiring from the position in 1981. A third brother (not pictured), Jack S. Liles Sr., also worked on the bridge-tunnel project for Tidewater Construction. He retired as president of the company in 1990 after 40 years with the company. (Courtesy of the *Eastern Shore News*.)

The two high-rise bridges on the Chesapeake Bay Bridge-Tunnel—the North Channel Bridge and the Fisherman Inlet Bridge—play a vital role by keeping traffic open for watercraft. The bridges are fixed-level structures with steel superstructure spans supported on piers. They are both 28 feet wide, with 18-inch emergency sidewalks on each side. Above, the North Channel Bridge was built at Cape Charles Harbor, where ferries once loaded and unloaded passengers. The bridge was actually built on a barge floating in the harbor, on top of scaffolding that raised it to its intended height on the Bay. When the bridge was finished, it was floated on the barge to the bridge-tunnel. As the tide dropped and ballast was pumped into the barges, the bridge was lowered into place. The first attempt to float the bridge was thwarted by thunderstorms, which drove the barge back to the safety of Cape Charles Harbor. (Courtesy of the *Eastern Shore News* and the Chesapeake Bay Bridge-Tunnel District.)

Above, bridge-tunnel commission chairman Lucius Kellam Jr., descending the ladder, inspects the North Channel Bridge as its deck is being built. The view below shows the North Channel Bridge with the main span having been placed. Altogether, the bridge is supported by 18 concrete piers. Three complete sets of construction rigs, each carrying a 75-ton crane, were needed to build the bridge piers in the North Channel. The North Channel Bridge features the highest point of the entire bridge-tunnel, 83.5 feet above the water's surface. The North Channel Bridge is the larger of the two high-level bridges. Pictured here, the North Channel Bridge is 3,793 feet long and has a 75-foot vertical clearance and a 311-foot horizontal clearance. (Both, courtesy of the Chesapeake Bay Bridge-Tunnel District.)

The approaches are built for the Fisherman Inlet Bridge. The bridge links Wise Point on Virginia's Eastern Shore to Fisherman Island, which is the southernmost of Virginia's barrier islands. Fisherman Inlet Bridge is owned by the Eastern Shore of Virginia Wildlife Refuge. This is the shortest component of the Chesapeake Bay Bridge-Tunnel complex. Its 457-foot span allows access to the Intracoastal Waterway, located at the southern tip of the Eastern Shore. The bridge provides a 40-foot vertical clearance at high tide and 110 feet of horizontal clearance. (Both, courtesy of the *Eastern Shore News*.)

A form is assembled for casting one of the pile caps of the Fisherman Inlet Bridge. The bridge's spans are supported on these five-pile bents. One crucial piece of equipment on the project was a floating central-mix concrete plant. With central mix, the components of concrete—including water—are combined, then delivered to the job site. With ready mix, water is mixed in trucks during transport. Barges delivered stone, sand, and cement to the floating plant. (Courtesy of the Chesapeake Bay Bridge-Tunnel District.)

A small barge places a dump truck on the north end of the Fisherman Island Causeway. The causeway is the earth-filled embankment 15 feet above mean low water that links the Fisherman Inlet Bridge and the North Channel Bridge. The road is 24 feet wide, with a 10-foot shoulder on each side. The side slopes of the embankment are protected from wave action and erosion by a blanket of stone riprap. (Courtesy of the Chesapeake Bay Bridge-Tunnel District.)

The chosen path for the bridge-tunnel crosses two of the busiest shipping channels in the world. On the Norfolk side, the Thimble Shoal Channel was, and still is, used heavily by the Navy, and the military would not hear of blocking the channel with a bridge. To the Navy, that meant the potential for the enemy to bomb the span and block the Navy's access to the ocean. On the north side is the Chesapeake Channel, used for ocean shipping. Maryland officials similarly objected to blocking that channel to Baltimore's port. The answer was the construction of two mile-long tunnels along the 18-mile span. The tunnels' odyssey began in Orange, Texas, at American Bridge Division of United States Steel Corporation. Above, tunnel sections are built at American Bridge. Below, a tunnel section is pictured just before it heads to Norfolk. (Both, courtesy of the Chesapeake Bay Bridge-Tunnel District.)

The last tunnel section leaves American Bridge in Orange, Texas, where the sections were fabricated, in September 1961. The 1,700-mile trip to Norfolk did not go without incident. Hurricane Carla sent two tunnel sections adrift the Gulf of Mexico. They were recovered days later, one after it washed up on a beach in front of a luxury hotel in Galveston, Texas. (Courtesy of the *Eastern Shore News*.)

With the help of tugboats, the tunnel section numbered "T-1" makes its way to the Sewells Point staging yard in Norfolk. In Texas, two tubes were built. One, an octagonal, double-walled form, is the outer shell; the other, a round tube with a reinforced steel webbing, is the tunnel. The design was referred to by engineers as "a can within a can." The tubes were built with this inner and outer core to allow the pouring of 4,000 tons of concrete between the shells. That concrete would play a crucial role in lowering the tubes into the Chesapeake Bay. (Courtesy of the Chesapeake Bay Bridge-Tunnel District.)

Above, a tunnel section is on its way to the "shape up" basin, or staging yard, at Sewells Point in Norfolk. Below, a section arrives at Sewells Point. The octagonal section is 37 feet wide. (Both, courtesy of the Chesapeake Bay Bridge-Tunnel District.)

At the outfitting pier in Sewells Point, in addition to the pouring of the inner concrete ring and roadway slabs, each section was also equipped with pipelines to supply and drain water, conduits for power and communications, ventilation ducts and flues, and electrical boxes and outlets. Each of the giant tubes is 286 feet long—about equal to a football field or a 20-story building. The interior is as wide and as high as a three-story building; 37 sections were required to form the two tunnels. (Courtesy of the Chesapeake Bay Bridge-Tunnel District.)

Chesapeake Bay Bridge-Tunnel Commission chairman Lucius J. Kellam Jr., left, and executive director J. Clyde Morris inspect progress at the Sewells Point fitting yard in 1961. Morris, who had been city manager of Warwick County, was named the bridge-tunnel's executive director in October 1957. One of his first acts was signing a contract for test borings along the newly approved route. (Both, courtesy of the Chesapeake Bay Bridge-Tunnel District.)

In this photograph, officials, including Percy Z. Michener (left), Leon R. Johnson (second from left), and J. Clyde Morris (far right), witness a tunnel section being moved across the bay. Michener was a project engineer for Sverdrup & Parcel, Johnson was chief engineer for the bridge-tunnel, and Morris was executive director of the Chesapeake Bay Bridge-Tunnel District. Michener joined the project in 1954, from the moment Sverdrup & Parcel sent him from its Washington office to help determine whether a crossing was feasible. (Courtesy of the Chesapeake Bay Bridge-Tunnel District.)

Above, the tunnel sections can be seen in various stages of outfitting, as represented by their respective elevations in the water. At right, the huge shell of a tunnel section is moored at the outfitting pier. Once outfitting was completed and the section was ready, all hatches would be sealed. Each segment of the tunnel is identical in design, though the Thimble Shoal Tunnel is made of 19 sections and the Chesapeake Channel Tunnel has 18 sections. (Both, courtesy of the *Eastern Shore News*.)

Above is a tunnel section as it appeared at the outfitting yard following its arrival from Texas. The circular, reinforcing steel gave the interior structural concrete ring its strength against the tremendous pressure of the seawater. Before its arrival at Sewells Point, the only concrete in the section was between the interior and exterior steel shells in the bottom of the tunnel section. That concrete provided stability during the tow from Orange, Texas. Below, the roadway slab was poured after the first stages of the structural ring were completed. (Both, courtesy of the *Eastern Shore News*.)

The reinforced concrete interior was poured in five equal segments. A lower section of the ring was poured, then haunches, then the roadway, then the walls, and finally the upper arch. The rolling form, below, shows the arch tying into the walls. The length of the form demonstrates one-fifth of the length of a tube segment. As the tunnel section at this stage was floating at the pier, the staging was considered carefully to avoid structural damage. When all the concrete was poured, the tube section would float about six inches above the waterline. (Both, courtesy of the Chesapeake Bay Bridge-Tunnel District.)

In the Bay, a large clamshell dredge, the *Virginian* (not pictured) had excavated a U-shaped trench to receive the tunnel section. A two-foot-thick foundation of sand and gravel was then placed and sloped to receive the tunnel section. Here, the first section has been placed at the island prior to its construction and is protruding above the water level. The next tunnel section is being moved into position to connect to the first. The first section has concrete blocks acting as ballast. All the tunnels had at least 10 feet of backfill, and—near the islands—heavy armor rock. All this weight served a purpose. It was intended to protect against floatation, even if the backfill were to be eroded by tidal currents. (Courtesy of the Chesapeake Bay Bridge-Tunnel District.)

At left, a diver descends from the placing barge to direct the connection of the tunnel sections. Divers worked in water depths up to 110 feet to assist in joining the sections. Joints were then sealed with underwater concrete, called "tremie." This was necessary before the bulkheads—the walls that sealed the ends of the tunnel sections—were removed. Powerful currents and waves made placing difficult, especially near the islands. (Courtesy of the Chesapeake Bay Bridge-Tunnel District.)

At right, the placing barge straddles a tunnel section as it is lowered into position. Ballast concrete was directed into the exterior pockets of the tunnel to increase its weight and stability. At this stage, the joints where the tunnel sections met were shored up with tremie—underwater concrete—to allow for removal of the bulkheads that segmented the tubes. Once the bulkheads were removed, workers could move freely between tunnel sections to complete interior work. Portal to portal, the Thimble Shoal Tunnel is 6,200 feet, and the Chesapeake Channel Tunnel is 5,664 feet. (Right, courtesy of Donna Sanders Corbus; below, courtesy of the *Eastern Shore News*.)

The walls and ceiling of the finished tunnel were covered with 5.5 million ceramic tiles. Water for this and other operations was provided from a 1,200-foot-deep well located on the South Island of the Thimble Shoal Tunnel. Finish work included the ventilation, lighting, power, and communications systems. Since the bulkheads nearest the islands could not be removed until island and ventilation building construction was complete, the tunnel sections nearest both islands had a "snorkel" shaft for access. Workers and materials were lowered 45 feet to the base of the tunnel from this shaft. A dismantled concrete truck was also lowered. (Both, courtesy of the Chesapeake Bay Bridge-Tunnel District.)

The two-lane roadway inside the tunnel was surfaced with asphalt, sloped to drain along both sides of the roadway. The drains had to accommodate water from washing and fire hoses. The visible interior work included hand railings, fire and telephone stations, and lighting. The lighting was in two continuous lines of fluorescent light, with intensities varied by zones to help drivers' vision adjust progressively upon entering the tunnel and nearing its exits. (Both, courtesy of the Chesapeake Bay Bridge-Tunnel District.)

Each tunnel has a roadway width of 24 feet between curbs and an overhead clearance above the roadway of 13.5 feet. At its lowest point, the roadway is 98 feet below mean low water. A maintenance catwalk is along one side of the tunnel. The tunnel ventilation is accomplished by a transverse distributional system that supplies fresh air uniformly along the tunnel length from a duct beneath the roadway and removes air through a duct above the ceiling. This photograph was taken in February 1964, just before the bridge-tunnel's opening. (Courtesy of the *Eastern Shore News*.)

One of the Chesapeake Bay Bridge-Tunnel's miracles was creating four islands in the middle of the Bay, where there had been nothing. The islands were created at each end of the two tunnels and are home to the ventilation buildings, tunnel portals, and approaches to connecting trestles. The islands were built by encircling the area with large stones—three abreast—then filling inside with sand pumped from the bottom of the Bay. This process was repeated—boulders, sand, boulders, sand, and so on—until the island surface was 30 feet above the water. Each island has an area of about eight acres. Above, the dredge *Ezra Sensibar* pumps sand to build one of the islands. The dredge, originally built in 1905, was refitted in 1961 for the tunnel project by replacing her steam engines with diesel engines. (Both, courtesy of the Chesapeake Bay Bridge-Tunnel District.)

All four islands are exposed to heavy sea conditions and storms. Once they were built to their final elevation, they were further protected with heavy armor rock weighing 10 to 20 tons, with seawalls around the top. In the above photograph, taken in September 1961, the survey tower mounted on the end tunnel section and the tunnel section itself can be seen protruding above the water surface. When this picture was taken, the island filling had not yet risen above the level of the sea. Below, the *Big D* is in the process of driving piles for the trestle bents, and Island No. 2, north of the Thimble Shoal Channel, is shown. The wave action around the island, acting against the protective armor stone, is clearly visible. (Both, courtesy of the *Eastern Shore News*.)

Lucius J. Kellam Jr., chairman of the Chesapeake Bay Bridge-Tunnel Commission, leads a team inspecting the south portal of the Chesapeake Bay Bridge-Tunnel. The sand was piled onto the north Chesapeake Channel island to help compress a peat bog sandwiched between layers of clay on the floor of the Bay. Before construction began, engineers and surveyors spent a year charting the exact layout of the project. A major consideration in determining the most economical alignment was the uneven Bay bottom. Its depth and soil conditions are the product of layers of sediment and drowned riverbeds. One hundred eleven borings were taken, some as deep at 310 feet. Each of the islands cost about $5 million to build. (Courtesy of the Chesapeake Bay Bridge-Tunnel District.)

The stones used to build the islands were so large that some arrived from quarries with only two per railcar. They weighed 10 to 14 tons each—but some were as much as 32 tons. It took 34,000 carloads of heavy boulders to hold the islands securely in place. The demand for $10 million in rock for the bridge-tunnel project proved a huge undertaking for Virginia rock producers. The bridge-tunnel would require 1,699,180 tons of rock and 500,000 tons of sand and gravel—in addition to materials for 185,000 cubic yards of ready-mix concrete. (Courtesy of the Chesapeake Bay Bridge-Tunnel District.)

As Hurricane Carla was wreaking havoc in Texas and Louisiana in September 1961, heavy swells on the night of September 7, 1961, caused the north island of the Thimble Shoal Channel to wash out. Gravel and quarry rock washed from the east rim of the island. Also, during the storm, a barge broke loose and struck the ferry *Old Point Comfort*. The islands were made to withstand hurricanes of 105-mile-per-hour winds. (Courtesy of the Chesapeake Bay Bridge-Tunnel District.)

"The islands were the biggest problem," Tidewater Construction Co. president Sam Liles would observe in May 1964, after the bridge-tunnel's opening. "We thought it would be feasible to use hydraulic dredges, but we found early we could not keep the floating pipelines intact because the rough water broke them loose and carried them to sea." (Courtesy of the Chesapeake Bay Bridge-Tunnel District.)

Man-made islands provide the entrance into the mile-long Thimble Shoal Channel, above. At the north edge of the Chesapeake Channel, below, builders discovered a soft peat bog trapped between two layers of clay, 70 feet below the Bay bottom. More than 3,500 vertical sand drains and 23 deep wells were driven into the Bay bottom to drain the bog. To compress the bog, thousands of tons of sand were piled on top of the island. As the pumps drained the water, the weight of the sand squeezed the peat into a thin, hard layer. It took the two-million-ton island about a year and a half to sink the requisite five to eight feet. (Above, courtesy of the *Eastern Shore News*; below, courtesy of the Chesapeake Bay Bridge-Tunnel District.)

Buildings located at both the southern (pictured) and northern terminuses of the Chesapeake Bay Bridge-Tunnel house both toll administration and control rooms for bridge-tunnel police. The photographs show the South Toll Plaza during and just after construction as part of the original span. Though the original toll booths on both sides of the bridge-tunnel would be demolished 35 years later, as part of the Parallel Crossing Project, the plaza buildings remained. (Both, courtesy of the Chesapeake Bay Bridge-Tunnel District.)

Each of the four man-made islands has a ventilation building connecting the tunnel approach ramp that leads to the trestle. The ventilation buildings each house six mammoth fans that pump air in and out of the mile-long tunnels under the shipping channels. At least two of each run at all times, their loud whirling sounds similar to airplane propellers. The fans replenish air inside the tunnels at a rate of 1,197,600 cubic feet per minute, cleaning the air from thousands of cars each day. They also help prevent buildup from moisture that could damage tunnel equipment. The ventilation buildings are nondescript, shielding the fact that they are each five stories tall. (Left, courtesy of the Chesapeake Bay Bridge-Tunnel District; below, courtesy of the *Eastern Shore News*.)

Electric power is supplied by a substation in Virginia Beach through two feeder cables. Each feeder cable has sufficient capability to operate the tunnel facilities. As an added safeguard, an engine-driven electric generator is also installed in each ventilation building to preserve continuity of services in case of emergency. The bridge-tunnel includes miles of electric cables and walls of power breakers and transformers—most housed to this day in their original cabinets. Each tunnel has eight industrial drainage pumps and concrete-encased water pipes. Little of this is visible to drivers. Each system or piece of equipment has at least one backup—attributable in part to the constant maintenance required in a salt-air environment. Workers known as "maintainers" check every piece of equipment on a two-to-three-hour schedule. (Both, courtesy of the Chesapeake Bay Bridge-Tunnel District.)

In the 1960s, the 4.38-mile four-lane approach road on the Eastern Shore end of the bridge-tunnel was built to interstate specifications in anticipation that the crossing may eventually become part of the interstate system. In the photograph above is the North Toll Plaza. The building closest to the bridge-tunnel is the maintenance building, and the blue-brick administration building is in the foreground. The facility is at Wise Point, on the site of a large mansion owned by the Wise family once known as Hallett Place of Cape Farm. Capt. Hugh Wise bought the property at auction in 1902 for $7,000. (Both, courtesy of the Chesapeake Bay Bridge-Tunnel District.)

IN MEMORIAM

HONORING THE MEN WHO LOST
THEIR LIVES BUILDING THE
CHESAPEAKE BAY
BRIDGE-TUNNEL

DONALD P. CARTWRIGHT
JOHNNIE DAVIS
HUBERT B. DYCUS
JOHN A. FEDOROWICZ
WILL A. SAWYER, JR.
JAMES A. WERNER

ERECTED JUNE 17, 1964

Seven workmen died building the original Chesapeake Bay Bridge-Tunnel, including one who died from a fall working on the tunnel sections in Orange, Texas. His name is not listed on this plaque. Of the others killed, one died by electrocution reaching for a wrench while he was standing in electrified water on one of the four islands; two died in the collapse of a crane boom; two died in a boiler explosion on a dredge; and one was fatally struck by a broken cable. By the time of the Parallel Crossing expansion more than 30 years later, construction safety was greatly improved, and no lives were lost during construction. This plaque hangs on the ventilation building on the south island at Thimble Shoal Channel. (Courtesy of the *Eastern Shore News*.)

On January 14, 1964, the first caravan of cars crossed the Chesapeake Bay Bridge-Tunnel. The last concrete had been poured on the bridge on January 10, and the concrete had just hardened by the fourth day to allow the crossing, though the bridge-tunnel would not open to the public for another three months. The group included contractors and engineers who had been involved in the project, as well as media representatives. The trip took 40 minutes and included a stop for photographs. After the caravan, the group stopped at a restaurant for lunch. One of the media representatives, Joel Carlson of WTAR-TV, proposed a new organization, the Order of the Soaring Seagull. Sam Liles, the president of Tidewater Construction, quipped that its motto would be "Look out below!" (Courtesy of the *Eastern Shore News*.)

Three

FROM PINES TO PALMS

On April 15, 1964, by the South Toll Plaza in Chesapeake Beach, the Chesapeake Bay Bridge-Tunnel was formally opened. Gov. Albertis S. Harrison, at the podium, addressed the gathering. Those in attendance included Bridge-Tunnel Commission chairman Lucius Kellam Jr., front row left, and Fred Duckworth, front row right, a commission member representing Norfolk. (Courtesy of the *Virginian-Pilot*.)

The Chesapeake Bay Bridge and Tunnel Commission

cordially invites you to attend

the Opening Ceremonies

for the

Chesapeake Bay Bridge-Tunnel

on Wednesday, April 15, 1964

at 2:00 p. m.

at the Southern Approach of the Bridge-Tunnel

in the City of Virginia Beach

R.S.V.P.

INVOCATION
The Right Reverend William A. Brown, D.D.
*Retired Bishop of the Diocese of Southern Virginia
of the Episcopal Church*

MASTER OF CEREMONIES
The Honorable Lucius J. Kellam, Chairman
Chesapeake Bay Bridge and Tunnel Commission

REMARKS
The Honorable Elbert N. Carvel
Governor of Delaware

The Honorable J. Millard Tawes
Governor of Maryland

The Honorable Harry F. Byrd, Sr.
United States Senator

INTRODUCTION OF SPEAKER
The Honorable W. Fred Duckworth, Vice-Chairman
Chesapeake Bay Bridge and Tunnel Commission

PRINCIPAL ADDRESS
The Honorable Albertis S. Harrison, Jr.
Governor of Virginia

DELIVERY OF PROJECT
Mr. S. E. Liles, President
Tidewater Construction Corporation

SPONSOR
Mrs. Lucius J. Kellam

BRIDGE-TUNNEL TOUR
Motorcade of Officials and Guests
Proceed Across the Chesapeake Bay Bridge-Tunnel
to North Toll Plaza and Return to Chesapeake Beach

Music Furnished by United States Navy Band

The original opening date was set for October 7, 1963. The date was changed to February 1, then April 15, 1964. Delays in construction were caused by stormy weather and a 26-day strike by an operating engineers' union. In his remarks, Governor Harrison stressed the fact the project had been built without state tax money and without federal aid. It was a project so large in scale, Harrison said, that many had believed only the federal government could handle it. "People in their own states can perform miracles," the governor concluded. Harrison predicted an era of growth for the two sections of Virginia linked by the bridge-tunnel—the pastoral Eastern Shore and the industrialized Hampton Roads region. The governor closed his remarks by referring to the Chesapeake Bay Bridge-Tunnel as "a necklace of pearls" gracing Virginia. (Both, courtesy of the Chesapeake Bay Bridge-Tunnel District.)

The final remarks at the opening were by Sam Liles, the president of Tidewater Construction and the chairman of the executive committee, Tidewater-Merritt-Raymond-Kiewit, Joint Venture Contractors. The entirety of his remarks stood in contrast to a project that was epic in scope: "On behalf of the contractors, I wish to state that the Chesapeake Bay Bridge-Tunnel crossing is now complete and ready for traffic. It is with the greatest of pleasure, therefore, that I turn this project over to you and the commission." Following those remarks, just before 3:00 p.m., Mr. and Mrs. Lucius Kellam Jr., along with Governor Harrison, cut the ribbon. Appropriately, 50 seagulls were released to signal the bridge-tunnel opening. A white seagull on a field of blue is the official roadside symbol for the bridge-tunnel. "Follow the sign of the soaring gull," early advertisements for the bridge-tunnel read. (Above, courtesy of the *Virginian-Pilot*; below, courtesy of the Chesapeake Bay Bridge-Tunnel District.)

The first crossing included a motorcade of 30 cars and 117 buses chartered for the occasion by the Chesapeake Bay Bridge-Tunnel Commission. The officials and guests were caught in a downpour for the round-trip tour of the new bridge-tunnel. Inclement weather, long the bridge-tunnel's builders' albatross, had visited again during the opening—as it would on April 15, 2014, when the bridge-tunnel marked its 50th anniversary. In 1961, George F. Ferris, chairman of Raymond International, observed: "In spite of all our advanced planning, our performance depends on average good luck with the weather." (Both, courtesy of the Chesapeake Bay Bridge-Tunnel District.)

The first toll was paid by brothers Omero C. Catan and Michael Katen of New York. Omero had earned the nickname "Mr. First," a result of a lifelong pursuit of firsts. Catan paid the first toll on the George Washington Bridge in 1931. He was first to buy a token on the Eighth Avenue subway and the first paying customer to skate on the Rockefeller Plaza ice rink in 1936. In all, he registered 537 firsts in his life, and he was usually accompanied by his older brother. The fact that the younger brother laid claim to "first" eventually became a point of dissention between the brothers. In later years, they squabbled over who was the rightful owner of the Mr. First moniker. The first southbound crossing from Cape Charles was made by Mr. and Mrs. John N. Hendreckson of Burlington, New Jersey. (Courtesy of the Cape Charles Historical Society.)

The aerial view above features Wise Point, on Virginia's Eastern Shore, looking south. Wise Point is the northern terminus of the bridge-tunnel. The mainland of Virginia is on the far horizon, with the Atlantic Ocean on the left. It was widely believed that the greatest cultural impact from the bridge-tunnel would be felt on the Delmarva Peninsula—one of the most insular sections of the heavily populated East. "I know the people of the Eastern Shore," Gov. Albertis S. Harrison Jr. said in 1964, "and know that changes will fall within the confines of Virginia traditions." On the left, a car passes over the North Channel Bridge by Fisherman Island. (Both, courtesy of the *Eastern Shore News*.)

The southern approach is in the Chesapeake Beach community of Virginia Beach on US 13, a mile north of US 60 and just east of the Little Creek Amphibious Base. Hampton Roads—then popularly referred to as Tidewater—had long been limited in its contact with the outside world. While many celebrated the advent of the bridge-tunnel, some in the Chesapeake Beach community, Chic's Beach or Chick's Beach to locals, did not care for the newfound attention, as they had always enjoyed less public traffic due to restricted parking and little renown. An old drive-in movie theater is visible at left above. (Above, courtesy of the Cape Charles Historical Society ; below, courtesy of the *Eastern Shore News*.)

Toll-booth operators at the bridge-tunnel work as part of the operations division, which includes toll collections and police. There are a total of 150 employees at the bridge tunnel. Other divisions include maintenance, executive, and finance. Maintenance itself has four departments: shops and services, electronics-communications, electrical-mechanical, and special projects, which oversees maintenance and repairs projects that "are unusual in nature and scope." The finance department's responsibility transcends money matters into community relations. The bridge-tunnel maintains an active presence with civic organizations. (Courtesy of the Chesapeake Bay Bridge-Tunnel District.)

With rampant speculation about how the Chesapeake Bay Bridge-Tunnel would alter the dynamic of the Eastern Seaboard, it is not surprising that initial traffic estimates were optimistic. Initial projections were that 5,000 cars would use the bridge-tunnel daily in its first year. In fact, just more than 2,000 cars used the bridge-tunnel daily in its first year. It would take until the late 1970s before traffic counts reached the original projections. Totals continued a steep climb, and by the mid-1980s, more than two million vehicles crossed the span in a calendar year. By 2000, the total had reached three million. For the past 10 years, the annual crossing has been about 3.5 million vehicles. (Above, courtesy of the *Eastern Shore News*; below, courtesy of the *Virginian-Pilot*.)

Emergency telephones are located at quarter-mile intervals along the entire bridge-tunnel project. The bridge-tunnel has its own wrecker service—including one truck still in service after 50 years. When vehicles break down on the span, they are towed to one side or the other, and it is left to the motorist to call the towing service of his or her choice. (Courtesy of the Chesapeake Bay Bridge-Tunnel District.)

The bridge-tunnel commission turned to bona-fide "Mad men," that is, Madison Avenue advertising executives, for help in publicizing the new span. They procured the services of the New York public relations firm Norman A. Schorr & Associates. Weekly construction updates were issued, and newsrooms were given special tours of the span. Films were even produced for distribution to television and movie theaters. And millions of pages of printed material were distributed, including brochures and maps. (Courtesy of the Cape Charles Historical Society.)

America House motel was built near the North Toll Plaza in 1965, just after the opening of Chesapeake Bay Bridge-Tunnel. "You can't go from 2,000 to 5,000 cars a day without creating a lot of additional business," observed Bridge-Tunnel Commission chairman Lucius Kellam in 1964. "This traffic increase will create restaurants, motels and service stations, which will contribute to allied industries." The hotel still stands today, following several name changes. (Courtesy of the *Eastern Shore News*.)

Sea Gull Island, the southernmost of four man-made islands, was built as the public face of the bridge-tunnel's four man-made islands. It is a recognition of the fact that all of the fascination surrounding construction of the span meant it would be something more than an entity that facilitated tourism: it would be a tourism destination in its own right. Sea Gull Island offers a chance to grab something to eat, fish, bird watch, or just take in the view of the bridge-tunnel. Each of the four islands can accommodate about 250 parked cars. (Both, courtesy of the Chesapeake Bay Bridge-Tunnel District.)

The bridge-tunnel's four islands offer opportunities for watching ships as well as for birding enthusiasts. The habitat of concrete, steel, asphalt, and rocks in the midst of open water makes for a hard-to-resist rest stop for birds during spring and fall migrations. Among the birds and waterfowl commonly seen on the bridge-tunnel's man-made islands are northern gannet, brant, harlequin duck, peregrine falcon, little gull, American white pelican, king eider, red-breasted merganser, American oystercatcher, and black-tailed gull. Heightened security went into effect post-9/11, but the bridge-tunnel makes special opportunities available for bird-watching groups to visit the islands that have restricted access. Groups of not more than 15 birders pay a fee to cover costs, including the police escort that remains with the group. Scientists and researchers may also take advantage of the provisions for special access. (Courtesy of the *Eastern Shore News*.)

In its first decade, the restaurant on the first northbound island—3.5 miles from the Virginia Beach shoreline—was a coffee shop and snack bar serving "ocean-fresh" fish and hamburgers and hot dogs. In the 1970s, the snack bar evolved into the Sea Gull Pier Restaurant & Gift Shop, known for its (seemingly) foot-long flounder sandwiches and free drink coupons. It was also known for its dark wood paneling. A 2010 renovation added more windows and a fresh Bay-appropriate color scheme. Today, the building houses Chesapeake Grill, a casual full-service restaurant, and Virginia Originals, which offers Virginia-centric gifts, including gourmet foods and wine, children's items, clothing, customized gift baskets, and souvenirs. The building also has a "grab 'n' go" convenience area where snacks and drinks are sold. (Above, courtesy of the Chesapeake Bay Bridge-Tunnel District; below, courtesy of the *Eastern Shore News*.)

Sea Gull Island offers great viewing opportunities of the Chesapeake Bay, including of the naval ships that pass through the Thimble Shoal Channel. In recognition of the US Navy's presence in Hampton Roads—and the frequent sightings of its vessels from the island—the Hampton Roads Naval Museum partnered with the bridge-tunnel on a series of outdoor exhibits. The six exhibit panels highlight the history of the Navy on the Chesapeake Bay. The Hampton Roads Naval Museum is housed inside of the Nauticus museum on the Norfolk waterfront, less than 30 minutes from the southernmost terminus of the bridge-tunnel. The battleship USS *Wisconsin* is moored at the museum. (Above, courtesy of the Chesapeake Bay Bridge-Tunnel District; below, courtesy of the *Eastern Shore News*.)

Sea Gull Island, the first island northbound motorists reach, 3.5 miles after leaving Virginia Beach, has a 625-foot pier. There, free fishing is offered. The wheelchair-accessible fishing pier offers a deep-sea fishing experience sans boat. There are cleaning stations and a weighing station. There is no additional fee to use the pier. Though a fishing license is not required, fishermen must register with the Virginia Fisherman Identification Program every year. (Both, courtesy of the *Eastern Shore News*.)

Among the species one can catch from the pier are spots, grunts, rockfish, croakers, flounders, bluefish, drums, tautogs, sheepsheads, small sharks, and less appealing catches such as stingrays and the oyster toadfish, a muddy-looking scaleless fish. The water is too deep for crabbing. In the evenings, the lights on the pier attract fish. Its effectiveness as a fishing spot is represented in the summers, when the pier is sometimes elbow-to-elbow. (Courtesy of the *Eastern Shore News*.)

In addition to birds and waterfowl, the bridge-tunnel's islands have been home to other wildlife. Each year, in the late fall and early winter, seals visit the relatively quiet rocks of the bridge-tunnel. The water temperatures are to their liking, and they enjoy the quiet of three of the four islands, which require special permission for people to access. Because of less visibility, the harbor seals are more drawn to islands three and four. Even there, however, they are sometimes subject to the dangers of gill nets from fishermen aboard boats. Dolphins are also frequently spotted. (Above, courtesy of the Chesapeake Bay Bridge-Tunnel District; below, courtesy of Brian M. Lockwood.)

A charter-fishing cottage industry soon formed around the bridge-tunnel and its 12 miles of trestle. It is well known among fishermen that fish—including big ones—love to live around bridge pilings, which offer shade, cover, and strong current. At any given time, there are clusters of boats around the Chesapeake Bay Bridge-Tunnel's trestles. Some refer to the cluster as the armada, and at times, each piling on the span will seemingly have an accompanying boat. There are print and online fishing reports dedicated specifically to the Chesapeake Bay Bridge-Tunnel, and a great many photographs of prize catches of striped bass have been taken with the bridge-tunnel as a background. (Both, courtesy of *Eastern Shore News*.)

The bridge-tunnel garnered the interest of thousands of newspapers and magazines in all 50 states and throughout the world. Publications included *Fortune, Life, This Week, Business Week, Newsweek, US News & World Report, Ladies' Home Journal, McCalls, Saturday Evening Post*, and *National Geographic*. "Three gigantic machines, 150 to 180 feet long, are striding, rolling and swinging themselves across the mouth of the Chesapeake Bay," the March 1962 issue of *Popular Science*, left, read. " In assembly-line fashion, they're building a highway over virtually open sea . . . eliminating the last water barrier to the Ocean Highway [with] . . . an engineering wonder so bold in scale as to dwarf anything else of its kind in the world." *Reader's Digest*, below, named the span one of the five future wonders of the world, along with the Snowy Mountains Scheme in Australia, the Netherlands Delta Plan, the New York Narrows Bridge, and the Mont Blanc Tunnel between Italy and France, the longest vehicular tunnel. (Both, courtesy of the Chesapeake Bay Bridge-Tunnel District.)

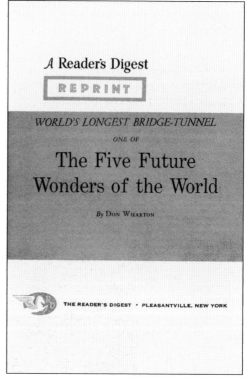

The American Society of Civil Engineers named the Chesapeake Bay Bridge-Tunnel one of the Seven Engineering Wonders of the Modern World after it opened in 1964 by virtue of its unusual engineering features, "usefulness to mankind," and size. The following year, 1965, the society awarded the bridge-tunnel its Outstanding Civil Engineering Achievement award. Among those who assembled to receive that honor were, at left, Sam Liles, president of Tidewater Construction, founded in 1932. Tidewater had also worked on the Midtown Tunnel and Hampton Roads Bridge-Tunnel. Second from left is George F. Ferris, chairman of the board for Raymond International, Inc., the world's largest builder of foundations for structures. At right, Leif J. Sverdrup, senior partner for the engineering firm Sverdrup & Parcel, founded his firm in 1928 and worked on the world's largest aerodynamic and proposal test facilities. Second from right, and pictured below, is Lucius Kellam Jr., chairman of the bridge-tunnel commission. The pictured plaque still hangs today in the lobby of the bridge-tunnel's administration building. (Both, courtesy of the Chesapeake Bay Bridge-Tunnel District.)

Lucius Kellam Jr. (right) received the first-place Virginia Travel Award in 1964 on behalf of the bridge-tunnel. The award was presented by James J. Geary (left), executive director of the Virginia Civil War Commission, which was formed to promote tourism during the Civil War centennial. The travel council was charged with "fostering and promoting the travel business [so] that the South may become the number one destination for American and World Travel." Toward that end, the council in the 1950s and 1960s spent much time marketing the benefits of through traffic from New England to Florida. Kellam received many commendations on behalf of the span and some specific to his efforts. In 1965, he was named Virginia Chamber of Commerce Man of the Year. In 1989, he received the Unsung Virginian Award from the Glen Allen–based Virginia, Maryland & Delaware Association of Electric Cooperatives. (Courtesy of the Chesapeake Bay Bridge-Tunnel District.)

Leaving the urban hustle and beachfront condos of Hampton Roads and—25 minutes later—finding oneself on the Eastern Shore, with its tidal marsh grasses, roadside stands, and pastoral beauty, is something like passing through a portal. Driving over the span, water on each side, one eventually loses sight of either shoreline for several miles in even the slightest haze. At night, the bridge-tunnel's long chain of lights is complemented by ship lights, boat lanterns, and blinking buoys. (Courtesy of the *Eastern Shore News*.)

On January 21, 1970, the 459-foot amphibious naval cargo ship *Yancey* had been anchored to the west of a trestle near the Virginia Beach shoreline. Strong winds caused her to drag anchor and smash into the span, 3.5 miles north of the Virginia Beach shoreline, in close proximity to Sea Gull Island, the first island driving northbound. The ship's crew attempted to lash the vessel to the trestle. The crew, with bridge-tunnel chief engineer Leon R. Johnson, was on the bridge deck when the ship struck the west side of the trestle. "I decided it was neither the time of day nor the sort of circumstances in which I wanted to hang around," Johnson would remark. Officials later surmised that the ship may have passed through a gap she created in the trestles, providing an opportunity to swing on her mooring line and hit the trestle again on its east side. (Courtesy of the *Eastern Shore News*.)

Three additional spans sustained crippling damage in the *Yancey* collision, including two girders and pilings, leaving a total of four 75-foot spans destroyed. It was the first substantial damage to the bridge-tunnel since December 3, 1967, when the *Mohawk*, a converted freighter, collided with the span. After the *Yancey* collision, divers went into the water and discovered that the 75-foot roadway sections were directly in the trestle path. It meant that all the trestle sections would have to be removed before new pilings could be driven and new sections set into place. The span was closed for a total of 42 days. (Both, courtesy of the *Eastern Shore News*.)

On September 21, 1972, the 235-foot Weeks No. 254 barge, which was empty, was being pulled in high winds by the tugboat *Carolyn*. The tug lost propulsion, her tow lines became entangled with the barge, and the two drifted together. The barge struck the west side of the bridge about 30 times over several hours before it became lodged under the weight of a collapsed portion of the bridge and beached near the bridge-tunnel's southern terminus. Damage was estimated at more than $1 million. Tourist and truck traffic shut down, Eastern Shore farmers could not reach their markets, and deliveries to Eastern Shore businesses such as hospitals and hotels ceased. As a stopgap measure, amphibious ferry service across the bay was reinstituted on a limited basis. (Both, courtesy of the *Eastern Shore News*.)

The Weeks barge closure caused much consternation, including in Virginia's capital, Richmond, and in Washington, DC, where the Navy Department was called into action. Even the US Secretary of Transportation became involved. Virginia governor Linwood Holton reportedly became agitated after he flew over the damaged facility and found the barge was still under the bridge. He urged haste in a conversation to then-Navy secretary (and later US senator) John Warner. After two weeks, the bridge-tunnel reopened to one-way traffic. Half a century later, during anniversary celebrations, the Weeks incident would be recalled as a ship dragging anchor drifted toward the bridge-tunnel in 80 mile-per-hour winds. Luckily, this ship did not strike the span. (Courtesy of the *Eastern Shore News*.)

Since opening in 1964, there have been a total of 78 fatalities related to traffic accidents on the Chesapeake Bay Bridge-Tunnel and a total of 51 fatal accidents. Most of the fatalities—72—occurred before the opening of the Parallel Crossing in 1999. The reason for the decrease is that, following the Parallel Crossing opening, there was no longer opposing traffic on the trestles. Traffic is only opposing inside the two one-mile-long tunnels. (Courtesy of the *Virginian-Pilot*.)

From its inception, the bridge-tunnel has had its own police force, commissioned under the same section of Virginia Code that applies to all political subdivisions in the commonwealth. Given its dynamic, as the sole funnel along the Atlantic coastline, it has been called "a big gill net across the road" or a "natural roadblock." Incidents can range from the mundane—breakdowns and traffic snarls—to traffic accidents, kidnappings, and murder investigations. Under "other duties as assigned," bridge-tunnel police are trained to handle "Chesapeake Bay Bridge-Tunnel Phobia," a marriage of "tunnel phobia" and gephyrophobia (fear of bridges). Occasionally, a driver freezes on the span. When that happens, about 10 times per month, an officer will drive the motorist's vehicle the rest of the way across. Drivers can even call ahead to request assistance at predesignated times. The original police force is pictured above in 1964. Below, the 1987 Chesapeake Bay Bridge-Tunnel District Police Academy poses for a photograph. (Both, courtesy of the Chesapeake Bay Bridge-Tunnel District.)

The Chesapeake Bay Bridge-Tunnel was celebrated from the outset as a faster direct route from southeastern Virginia to the Delmarva Peninsula, which includes Delaware and the Eastern Shore portions of Virginia and Maryland. The bridge-tunnel cuts 75 miles and 1.5 hours off the trip between Virginia Beach and Wilmington, Delaware. Prior to April 1964, motorists had to choose between a 90-minute ferry trip across the Chesapeake Bay or a long drive inland around the bay's upper reaches. Now they cross the mouth of the Bay in only 25 minutes. (Courtesy of the Chesapeake Bay Bridge-Tunnel District.)

To travelers' delight, tolls throughout the Chesapeake Bay Bridge-Tunnel's history have not kept pace with inflation. Though most travelers are familiar with only the rate for cars and light trucks, there are in fact rates assigned to 16 different categories of cars, heavy trucks, and buses. The one-way rate for a car was $4 in 1964. The toll climbed gradually through the years to its current rate of $13. The $10 rate implemented in 1991—concurrent with the Parallel Crossing opening—was raised to $12 in 2004. A glance at the posted toll rates in the photographs on this page sheds light on when the photographs were taken. The $6 posted toll, above, was in effect from 1975 to 1978, and an $8 posted toll, below, was in effect in 1978 and 1979. (Above, courtesy of the *Virginian-Pilot*; below, courtesy of the *Eastern Shore News*.)

The bridge-tunnel was named the Lucius J. Kellam Jr. Bridge-Tunnel in August 1987 to honor the man who moved the project from vision to reality. Virginia governor Gerald Baliles (left) was on hand. Kellam was chairman of the Chesapeake Bay Bridge-Tunnel Commission (originally the Chesapeake Bay Ferry Commission) from 1954 until 1993. When the original span opened in 1964, it was generally agreed that the bridge-tunnel would have remained a far-off dream had it not been for Kellam's relentless pursuit. To protect its name recognition, the facility continues to go by Chesapeake Bay Bridge-Tunnel. Asked in 1987 whether the span would now be known as the Lucius J. Kellam Jr. Bridge-Tunnel, Kellam himself remarked, "Probably not. But it's a big honor whatever they call it." After his death in 1995, Joint Resolution No. 28 was passed by the Virginia General Assembly, honoring Kellam as "a singularly dedicated and accomplished Virginian." (Both, courtesy of the *Eastern Shore News*.)

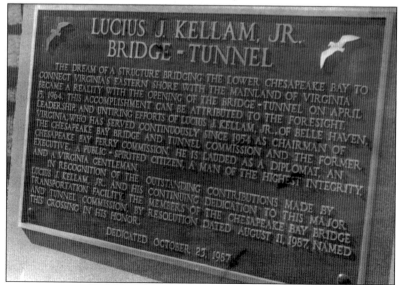

Four

THE PARALLEL CROSSING

Beginning in 1987, the Chesapeake Bay Bridge-Tunnel Commission started looking at the possibility of a Parallel Crossing. By 1989, studies showed such a crossing would be needed by 2000 to meet expected traffic volume, and the Virginia General Assembly gave the green light for the project in 1990. A joint venture of several out-of-state contractors was awarded the contract to build the second span west of and parallel to the original bridge-tunnel. They included PCL Civil Contractors, Inc., the Hardaway Company, and Interbeton, Inc. (Courtesy of the Chesapeake Bay Bridge-Tunnel District.)

The Parallel Crossing was started about 35 years after the original project and cost about $250 million. It was begun on June 16, 1995, and opened on April 19, 1999. On the Virginia Beach side of the span, above, contractors accustomed to battling the elements of the Bay faced a new challenge: how to make the water disappear. A portion of a small lake in Chesapeake Beach was effectively dammed to allow a pile driver to have access in order to build an approach bridge. After the parallel span was built, below, the fill material was removed. (Both, courtesy of Walter Grantz.)

With the original span, pile cutting would have been among the duties of No. 2 among the "Magnificent Machines" that built the bridge-tunnel, the *Two-Headed Monster*. For the Parallel Crossing, workers cut piles from a platform that included all three pilings in a bent, the term used to describe a grouping of three pilings. The pilings were cut to the correct height using jackhammers set on a track. The pilings pictured are in proximity to the new North Channel Bridge. (Courtesy of the Chesapeake Bay Bridge-Tunnel District.)

Building the high-rise North Channel Bridge, with its elevated approach spans, required heavier support structure than the typical 100-foot trestle spans. For this purpose, many cylinder piles were incorporated into these large "bathtub" structures. These structures, in turn, supported pier shafts of the varying heights needed to carry the approach spans as they rose to the main crossing of the North Channel. (Both, courtesy of the Chesapeake Bay Bridge-Tunnel District.)

When it came time to build a parallel North Channel Bridge, it was decided to build a steel plate-girder bridge rather than a truss-style bridge, like the original. The plate-girder bridge had all the strength minus the hassle that came with maintaining a superstructure. As part of the Parallel Crossing Project, the concrete roadbed of the original North Channel Bridge was replaced. Much of the Eastern Shore coastline today is protected, including the 1,123-acre Eastern Shore of Virginia National Wildlife Refuge. The refuge was established in 1984 to serve as an important stopover for migratory birds. Each fall, millions of songbirds and monarch butterflies converge on the tip of the Virginia Eastern Shore. (Right, courtesy of Walter Grantz; below, courtesy of the Chesapeake Bay Bridge-Tunnel District.)

The stars of the original Chesapeake Bay Bridge-Tunnel construction projects were the *Big D* pile driver, the *Two-Headed Monster*, which cut and capped bridge pilings, and the *Slab Setter*, which placed the road sections. For the Parallel Crossing, they were all replaced by a rig that had a name relatively lacking in pizzazz—IB909—but which made up for it in functionality. From the IB909, piles were driven and leveled, pile caps were set, and road decking was laid. The massive 1,800-ton barge could be jacked up on its four "spuds" safely above the surface of the water. (Courtesy of Walter Grantz.)

Built in the Netherlands in 1982, IB909 was called the world's largest "jack up" barge. It functioned something like a large ship, or a small city, in the middle of the bay. It included a mess hall, kitchen, cabins, bathrooms, offices, and a meeting room. It even had a facility called a tea room, perhaps a salvo to its European roots. The 12-mile low-level trestle forms the main component of the project, with precast and prestressed deck units. Like the originals, the pilings were 54 inches in diameter with precast caps. (Both, courtesy of Walter Grantz.)

For the Parallel Crossing, the concrete components were not produced at Bayshore Concrete in Cape Charles, which played such a crucial role in the initial span's construction. They were produced at a 27-acre waterfront precast yard near Little Creek in Virginia Beach, above. From there, the components were loaded onto barges with a gantry, below, then transported to the job site. At the job site, they would be loaded onto the IB909 with a crane. Whereas slabs were about 75 feet in the original project, the new prestressed concrete technology meant slabs could now be 100 feet long. (Both, courtesy of Walter Grantz.)

In a cast-in-place deck system, ready-mix concrete is placed into removable forms that are erected on site. First, rebar is placed on top of wood forms. Concrete is then poured on top of the rebar. Finally, an aluminum device called a "screed" is rolled over the surface of the road, leveling the concrete to create a smooth surface. Though precast concrete is less expensive and less labor intensive, cast-in-place concrete requires less maintenance. The Parallel Crossing's road is 38 feet wide, with a 10-foot-wide emergency shoulder on the right side. (Both, courtesy of the Chesapeake Bay Bridge-Tunnel District.)

The Chesapeake Bay Bridge-Tunnel Commission offered three primary reasons for the Parallel Crossing Project: to accommodate future traffic growth, to provide a safer crossing by eliminating most opposing traffic, and to allow for maintenance without completely closing off traffic. The project added 18 miles of new trestle, effectively creating a four-lane highway. But widening the tunnels was not part of the plan. This meant that the four lanes would have to become two where trestles meet tunnels. At the four-to-two-lane bottleneck, some widening of the islands was required. It meant importing more of the massive boulders that form the riprap around the perimeter of the islands. (Both, courtesy of Walter Grantz.)

This aerial view shows how the Parallel Crossing's trestles converge where they meet a two-lane tunnel. Ultimately, when the parallel tunnels are built, the islands will have to be widened to accommodate the second tubes. When the Parallel Crossing was built, engineers were already thinking about those new tunnels. With the exception of the northernmost two miles of trestle—where the new trestle is 500 feet from the original trestle—the new and old trestles are 250 feet apart. This is the distance needed to dig a parallel trench without compromising the safety of the first tunnel by disturbing the backfill and riprap that covers the tube. (Courtesy of Walter Grantz.)

As part of the Parallel Crossing Project, new toll booths were built. The two toll plaza buildings remained. When the Parallel Crossing was completed, the original span was closed for renovation, which included emergency pull-off areas. (Courtesy of the Chesapeake Bay Bridge-Tunnel District.)

The continuous cable trays alongside the bridge-tunnel keep cables from being suspended off the side of the bridge. The cables include electrical, signal, telephone, and fiber-optic lines. Pictured here is Walter Grantz, former chief engineer for the Chesapeake Bay Bridge-Tunnel. (Courtesy of Walter Grantz.)

The Chesapeake Bay Bridge-Tunnel complex is comprised of seven major component structures: 11.75 miles of low-level trestle; 1.75 miles of the Thimble Shoal Channel tunnel and islands; 1.5 miles of the Chesapeake Channel Tunnel and islands; .75 mile of the North Channel Bridge and approaches; .25 mile of the Fisherman Inlet Bridge and approaches; 1.75 miles of the Fisherman Island Causeway; and 5 miles of approach roads. The total length over water is 17.75 miles, and including the approach roads, it's 22.5 miles. The photograph above is a view from the Eastern Shore, or northern terminus, of the bridge-tunnel, showing Fisherman Inlet leading up to the North Channel Bridge. Below is a view of the Chesapeake Beach shoreline in Virginia Beach. (Both, courtesy of the Chesapeake Bay Bridge-Tunnel District.)

The bridge-tunnel is equipped with roadway lights across the entire span. Illumination is provided by 940 400-watt, 20,000-lumen mercury-vapor lamps mounted on prestressed, reinforced concrete poles. A study of the poles indicated that prestressed concrete would resist vibration in high winds, which means fewer bulb changes. The lamps are spaced 225 feet apart on alternating sides of the road. (Courtesy of the Chesapeake Bay Bridge-Tunnel District.)

The commission has started planning for a parallel tunnel adjacent to the northern Thimble Shoal Tunnel in 2021 at a cost of approximately $1.2 billion. It will begin its construction and environmental studies in 2018. The commission will finance this latest project through its reserves, bond sales, and low-interest loans. (Courtesy of the Chesapeake Bay Bridge-Tunnel District.)

On April 15, 2014, state senator Lynwood Lewis (above) addressed a gathering of more than 400 people at Sea Gull Island for a 50th-anniversary celebration. There was a special guest at the ceremonies—a vintage Willys Jeep. It was, possibly, the first automobile to drive across the Chesapeake Bay Bridge-Tunnel. Jeff Walker, today a member of the Chesapeake Bay Bridge-Tunnel Commission, bought the Jeep in 1966 from Welford O. Lucy Jr., an engineer with Sverdrup & Parcel who worked on the original crossing project. Welford—apparently determined to claim the title of first—told Walker he used boards to fill the gaps where there was no road yet. It had been mothballed at Walker's Nassawadox, Virginia, home since the late 1970s, but Walker, pictured below, decided to get it in running condition for the 50th anniversary. "It's kind of like it's come full circle," he said. (Both, courtesy of the Chesapeake Bay Bridge-Tunnel District.)

DISCOVER THOUSANDS OF LOCAL HISTORY BOOKS FEATURING MILLIONS OF VINTAGE IMAGES

Arcadia Publishing, the leading local history publisher in the United States, is committed to making history accessible and meaningful through publishing books that celebrate and preserve the heritage of America's people and places.

Find more books like this at
www.arcadiapublishing.com

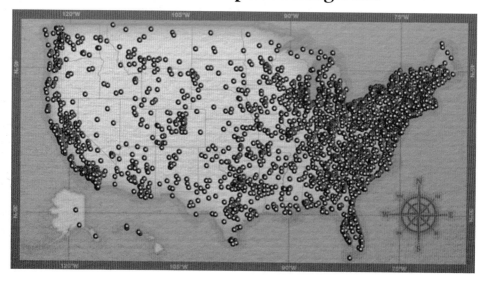

Search for your hometown history, your old stomping grounds, and even your favorite sports team.